T0135444

Scratch and Mar, Surface Structure and Rheology Measurements of Organic Coatings

A Brief Review

Michael Osterhold

Bibliografische Information der Deutschen Nationalbibliothek

Die Deutsche Nationalbibliothek verzeichnet diese Publikation in der Deutschen Nationalbibliografie; detaillierte bibliografische Daten sind im Internet über http://dnb.d-nb.de abrufbar.

© Copyright Logos Verlag Berlin GmbH 2019
Alle Rechte vorbehalten.

ISBN 978-3-8325-4870-4

Logos Verlag Berlin GmbH
Comeniushof, Gubener Str. 47,
10243 Berlin
Tel.:   +49 (0)30 42 85 10 90
Fax: +49 (0)30 42 85 10 92
INTERNET: http://www.logos-verlag.de

Scratch and Mar, Surface Structure and Rheology Measurements of Organic Coatings
A Brief Review

Michael Osterhold

Contents

Preface   5

Methods for Characterizing
Scratch/Mar Resistance   7

Measuring the Surface Structure
From Substrate to Topcoat   21

Improving Rheology Measurement
Yield Point and Thixotropy   33

Contributions in German

Methoden zur Charakterisierung
der Kratzbeständigkeit   43

Messung der Oberflächenstruktur
von Substrat bis Decklack   57

Verbesserung der Rheologie-Messung
Fließgrenze und Thixotropie   70

References   81

Preface

This short book contains three brief reviews written by myself covering recent topics in the field of 'physical characterization of organic coatings':

Scratch/Mar
Surface Structure
Rheology Measurement

The contributions have been published in the years 2016 to 2018 in the 'European Coatings Journal' (details see references). In addition, German versions of the articles can be found in the second half of the book.

Michael Osterhold, February 2019

Biography

Dr. Michael Osterhold studied physics at the Ruhr-University, Bochum (Germany), and received his Ph.D. in Experimental Physics in 1988. He then joined the Research Department of former Herberts Company (Hoechst Group), Wuppertal (Germany), since 1999 DuPont Perform-ance Coatings, where he was responsible for the unit R&D-Services EMEA (incl. Physics and Analytical). In 1995 he was recipient of the Farbe&Lack-Prize. He is author of about 70 scientific/technical papers. Since 2011 he has been working as a scientific consultant and as an adjunct Professor of Physics at an University of Applied Science.

Methods for Characterizing Scratch/Mar Resistance

Abstract

Scratch/mar is one of the most important physical properties to characterize the mechanical quality of a coating system. In this paper the relevant measuring techniques are briefly reviewed reflecting the application in the automotive and coatings industry. Techniques to determine the scratch/mar resistance based on single scratch tests are compared to other more simple methods as for example the car wash brush test (Amtec).

1 Introduction

In addition to good levelling, high gloss and effect development the resistance to mechanical damage – by stone chippings and scratching – is particularly important in order to obtain a high-quality appearance for clearcoats. The brushes and dirt in a car wash, for example, produce scratches measuring only a few micrometers in width and up to several hundred nanometers in depth.

With this background several measuring methods have been discussed over the last 15 to 20 years in the automotive and paint industry. The objective was to obtain a clear characterization of the scratch/mar resistance of clearcoats. Procedures that create a single scratch have been developed and improved at the end of the 1990s (micro or nano scratch method) [1-8]. These methods are different from more practically oriented procedures that are based on relatively simple methods to try to test or even come close to reality (e.g. car wash brush method).

2 Characterization Methods

2.1 Car Wash Brush Methods (Amtec)

The clearcoat to be tested (applied on a standard metal sheet) is moved back and forth 10 times under a rotating car wash brush. The brush (PE) is sprayed with washing water during the cleaning procedure. Because the metal sheets used for the test are clean, a defined amount of quartz powder is added to the washing water as a replacement for street dirt (Amtec method). A gloss measurement e.g. in 20°-geometry is used to evaluate the scratch resistance. The initial gloss and the gloss after the cleaning procedure are measured. The percentage of residual gloss with regard to the initial gloss is a measure for scratch resistance. High values indicate good scratch resistance, the instrument is schematically shown in fig. 1 [8]. A commercial car wash brush instrument is manufactured by Amtec-Kistler/Germany.

This method was developed in a project working group of DFO (Deutsche Forschungsgesellschaft für Oberflächenbehandlung) two decades ago and has been specified in the standard DIN EN ISO 20566. More details can be found for example in [9].

2.2 Crockmeter

The crockmeter is used by large car manufacturers and represents a different type of strain. The crockmeter has been the standard device of the American Association of Textile Chemists and Colorists (AATCC). It has been mainly used to test textiles for color fastness and abrasion.

This instrument (Atlas Materials Testing) is equipped with an electrical motor, so that a uniform stroke rate of 60 double strokes/min is reached. The sample is fixed on a flat pedestal. The sample is exposed to linear rubbing caused by a cylinder, other geometries are described in DIN 55654. A special testing material is attached to the bottom side of the cylinder, which is 16 mm in diameter; its downward force is 9 N. Ten double strokes are carried out over a length of 100 mm. For this method, the percentage of residual gloss is also used as a measure for scratching.

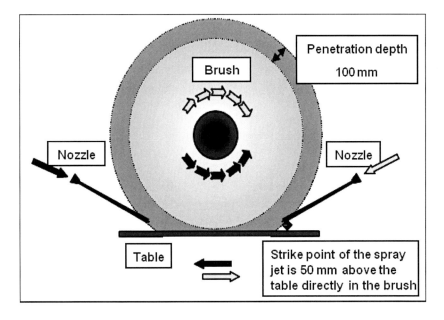

Fig. 1: Car wash brush method (Amtec)

In [8] a good correlation of both methods, crockmeter and laboratory car wash, in the range of good and bad scratch resistance can be observed. For the range of medium scratch resistance there is no clear correlation detectable. The reason for this could be the different degrees of strain in the two scratch test methods. The Amtec test is the one with a rather higher strain load.

2.3 Rota-Hub method

The Rota-Hub-Scratch-Tester has been developed by Bayer AG approx. 15 years ago . In this test, a carriage moves in x- and y-directions. A rotating disc with the scratching medium (e.g. paper) is attached to the carriage. The rotating disc is lowered onto the sample. This way the sample is strained by a rotating disc that also moves in x- and y-directions. Feed rate, rotation velocity, and contact pressure can be randomly selected and have been optimized.

For the automotive industry copy paper has proved to be a suitable scratching material because it causes scratches similar to those caused by a car wash. The resulting damage is a meander-shaped scratch pattern that can be measured using the parameters gloss and haze [10].

2.4 Micro-scratch experiments

In micro-scratch experiments single scratches of characteristic and realistic phenotype can be generated. The indentation depth depends on the applied force and the indentation body (indentor). Indentation depth is usually in the range of 1 µm and smaller. A diamond indentor (radius at the peak 1 – 2 µm) is pushed onto the sample surface applying a defined force to generate the scratch while the sample is moved in a linear direction at constant velocity underneath the indentor (see fig. 2). The applied normal force can be constant or progressive during the scratching procedure.

Basically it is possible to measure the tangential forces and the indentation depth during the scratching procedure. Deformations or damages can be observed with a microscope or additionally with a video camera. The profile of the generated scratches can be measured using AFM technology. In combination with the parameters measured during the scratching procedure it is possible to calculate physical parameters which allow conclusions about the elastic and plastic deformation behavior and the fracture behavior.

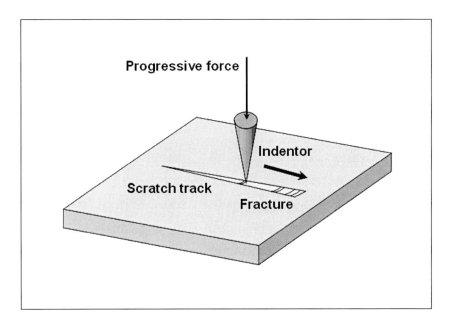

Fig. 2: The principle of single-scratch tests

2.4.1 Tests with progressive load

In this context a method was developed in the DuPont Marshall Lab. (Philadelphia) that generates and evaluates a single scratch on a surface (nano-scratch method). This method has been used for tests along with development work for clearcoats.

A commercial instrument by the Swiss company CSM (now part of Anton Paar company) was tested in the course of a project at the Research Institute for Pigments and Paints (Forschungsinstitut für Pigmente und Lacke e.V. (FPL), now part of IPA/Stuttgart) around the year 2000 and later. This instrument is based on the method developed at the DuPont Marshall Lab. Objective of several projects – where manufacturers of paints, raw materials and automobiles worked together – was the evaluation of the nano-scratch-tester regarding reproducibility, accuracy and applicability under the aspect of a realistic determination of scratch resistance.

The key point in this method is the determination of the critical load where first irreversible cracks or fractures are generated and which therefore indicates the transition from plastic deformation to significant/lasting damages. For that, the normal load is constantly increased and indentation depth and tangential load are simultaneously recorded. The transition from plastic deformation to the fracture range is indicated e.g. by unsteadiness or fluctuations in the detected load flow and the indentation depth. As a typical dimension besides the critical load it is common to determine the residual indentation depth after scratching – typically at a normal load of 5 mN – in the range of plastic deformation.

In figure 3 the transition range of an acrylate/melamine system is demonstrated. The top picture shows an AFM image of the scratch ridge itself where cracks are clearly visible. The bottom picture is taken with a microscope at a magnification factor of app. 1000, see also [8, 11, 12].

If values from nano-scratch tests are compared to results from car wash tests (see fig. 4 [8]), a certain correlation to the critical load can be observed. From this it can be concluded that the Amtec test rather causes a significant damage than strains in the plastic range [13].

3 Dynamic Mechanical Analysis

For certain clearcoat systems a partial healing of scratches can be observed on the time scale. In literature this is known as the reflow effect [14]. Thermal relaxation phenomena may be used for a physical explanation of this effect. In connection with scratch/mar resistance the cross-linking density of clearcoats is also a decisive factor. Dynamic mechanical analysis (DMA) has been established as a method to determine cross-linking density [14-16].

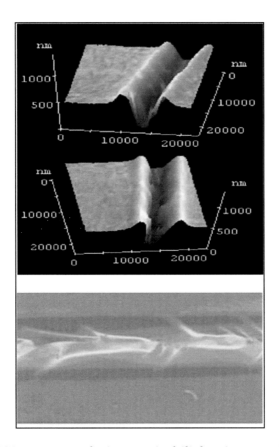

Fig. 3: AFM image top and microscopical (light microscope) image (bottom) of cracks in an acrylate/melamine system

Several ranges can be characterized when observing the behavior of the storage modulus E´ as a function of temperature. The glass transition range is followed by the rubber elastic range, which can be more or less distinctly depending on the degree of cross-linking. In this range the value of the storage module E´ is connected with the cross-linking density. The value of E´ at the local minimum is often chosen as a quantitative measure for cross-linking density.

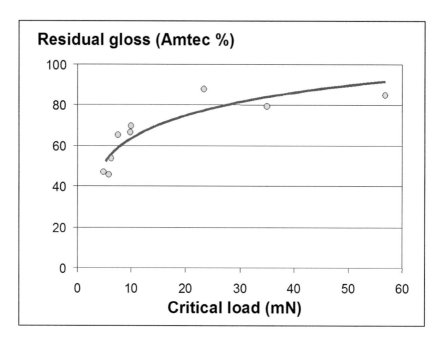

Fig. 4: Comparison of values from Amtec measurements to values for critical load from nano-scratch tests

The clearcoats were examined by submitting free films to a tension test. A DMA 7 instrument by Perkin Elmer was used for these tests. In this context, measurements of paint systems are described e.g. in [16].

Figure 5 shows a comparison of the results of DMA analysis and values obtained in car wash simulations [8].

A clear correlation between E´ as a measure for cross-linking density and the value for scratching can be observed. The quality of the scratching level (high residual gloss) increases with increasing cross-linking density [17].

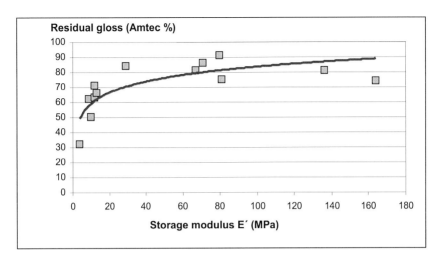

Fig. 5: Correlation of scratch resistance (Amtec values) and crosslinking density (storage modulus E´)

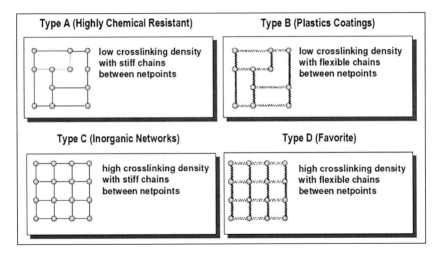

Fig. 6: Different network types

Fig. 6 shows a short comparison of different network types and brief remarks of their properties [18, 19, 20]; with type D as a favorite for automotive clearcoats.

4 Influence of weathering

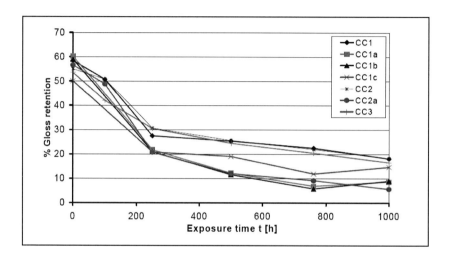

Fig. 7: Gloss retention (Amtec) troughout SAEJ1960 weathering

The influence of weathering on scratch resistance is demonstrated in fig. 7, showing the gloss retention after Amtec testing of various clear-coats as a function of exposure time of SAEJ1960 weathering. All samples that were investigated show a significant drop of residual gloss after a short exposure time. A clear differentiation between the clear-coats with (CC1, CC2, CC3, CC1c) and without light-screener (CC1a, CC1b, CC2a) is noticeable. Afterwards the residual gloss decreases slightly [21, 22]. Other studies can be found e.g. in [23].

5 Summary

In this paper, several methods to determine the mar (scratch) resistance of clearcoats are presented. Methods that create a single scratch are different from more practically oriented procedures that are based on relatively simple methods to try to test or even come close to reality (e.g. car wash brush method). The methods were reviewed briefly considering also other physical properties like cross-linking density or weathering.

Recent investigations show that scratch/mar resistance is always an important factor to characterize physical properties of coating surfaces, for example in analyzing automotive clearcoats containing silane modified blocked isocyanates [24]. Their results showed that a close correlation existed between the scratch resistance data obtained from car-wash and nano-scratch tests for certain clearcoats.

Also, scratch resistance of exterior clearcoats and polycarbonate hardcoats were examined and discussed recently by [25].

Studying the literature there is often no clear separation between the expressions mar and scratch resistance. Mar should refer to light surface damages encountered in the real field that are usually shallow and narrow while scratch refers to medium or more severe damages [26]. However, several authors handle this uncertainty in writing scratch/mar.

References

01.   G. S. Blackmann, L. Lin, R. R. Matheson, ACS Symposium Series: Microstructure and Tribology of Polymer Surfaces (1999)

02.   L. Lin, G. S. Blackman, R. R. Matheson, Prog. Org. Coat., 40 (2000) 85

03.   K. Adamsons, G. Blackmann, B. Gregorovich, L. Lin, R. Matheson, Prog. Org. Coat., 34 (1998) 64

04.   J. L. Courter, Proc. 5th Nürnberg Congress, Nuremberg, Germany, (1999) 351

05.   E. Klinke, C. D. Eisenbach, Proc. 6th Nürnberg Congress, Nuremberg, Germany, (2001) 249

06.   L. Lin, G. S. Blackman, R. R. Matheson, Materials Science and Engineering, A317 (2001) 163

07.   G. Wagner, M. Osterhold, Mat.-wiss. u. Werkstofftech., 30 (1999) 617

08.   M. Osterhold, G. Wagner, Prog. Org. Coat., 45 (2002) 365

09.   E. Fischle, Farbe Lack, (2009), No. 04

10.   E. Klinke, M. Kordisch, C.D. Eisenbach, T. Klimmasch, Farbe Lack, 108 (2002), No. 2, 55

11.   M. Osterhold, Europ. Coat. Journal, (2005), No. 09, 34

12.   M. Osterhold, Progr. Colloid Polym. Sci., 132 (2006) 41

13.   K.-F. Dössel, SURCAR, Cannes, France, 2001

14.   R. Gräwe, W. Schlesing, M. Osterhold, C. Flosbach, H.-J. Adler, Europ. Coat. Journal, (1999), No. 3, 80

15.   W. Schlesing, Farbe Lack, 99 (1993) 918

16.   W. Schlesing, M. Buhk, M. Osterhold, Prog. Org. Coat., 49 (2004) 197

17.   E. Frigge, Farbe Lack, 106 (2000), No. 7, 78

18. M. Osterhold, Vortragstagung der GDCh-Fachgruppe Anstrichstoffe und Pigmente, Eisenach, Germany, 2005

19. C. Flosbach, XXVIII Fatipec Congress, Paper II.C-1, Budapest, Hungary (2006)

20. K.-F. Dössel, in: Automotive Paints and Coatings, 2nd Ed., H.-J. Streitberger, K.-F. Dössel (Eds.), Wiley-VCH (2008)

21. M. Osterhold, B. Bannert, W. Schubert, T. Brock, Macromol. Symp. 187, (2002) 823

22. B. Bannert, M. Osterhold, W. Schubert, T. Brock, Europ. Coat. Journal, (2001), No. 11, 30

23. C. Seubert, M. Nichols, K. Henderson, M. Mechtes, T. Klimmasch, T. Pohl, J. Coat. Technol. Res., 7 (2010), 159

24. S. M. Noh, J. W. Lee, J. H. Nam, J. M. Park, H. W. Jung, Prog. Org. Coat., 74 (2012) 192

25. C. Seubert, K. Nietering, M. Nichols, R. Wykoff, S. Bollin, Coatings, 2 (2012) 221

26. Z. Ranjbar, S. Rastegar, Prog. Org. Coat., 64 (2009) 387

Measuring the Surface Structure
From Substrate to Topcoat

Abstract

Besides color, effect and gloss, the visual impression of a painted surface is influenced especially by the surface structure (leveling, waviness, orange peel, appearance). To characterize the structure of a surface, mainly two different measuring methods – profilometry and 'wave-scan' – have been established in the automotive and paint industry. The mechanical profilometry in combination with Fourier techniques allows detailed investigations concerning the surface topography from substrate to topcoat. For example, influences of the substrate on the different coating layers and other effects can be evaluated. Basic relations and relevant application examples from the areas 'metal, plastics, coating', investigated with profilometry and 'wave-scan' in the course of time, will be summarized in this paper. Results obtained from coating surfaces with lower gloss will also be presented.

1 Introduction

In addition to color, effect and gloss, the visual impression of a painted surface is influenced especially by the surface structure (leveling, waviness, orange peel, appearance). To characterize the structure of a surface, mainly two different measuring methods – profilometry and 'wave-scan' – have been established in the automotive and paint industry. The mechanical profilometry combined with Fourier techniques

(FFT) yields detailed information of the surface topography, and substrate influences or other effects on the final coating appearance can be described [1-18]. To simulate the visual impression obtained from optical inspection of surface structures, the German company Byk-Gardner developed the so-called 'wave-scan' instruments. In addition to measurements on glossy surfaces, i.e. topcoats/clearcoats, the 'wave-scan *dual*' allows to detect the appearance of surfaces with lower gloss (medium glossy), e.g. primer-surfacers and sometimes EC. Basic relations and relevant application examples from the areas 'metal, plastics, coating', investigated with profilometry and 'wave-scan' in the course of time, will be summarized in this paper.

2 Methods

2.1 Mechanical surface characterization

Surface profiles presented in this article were measured by mechanical profilometry using the Hommeltester T 8000 (Hommel-Etamic, Germany). For all measurements, a dual-skid tracing system or a so called datum system (without skids) with a diamond tip radius of 5 μm was used. The vertical resolution of this mechanical profilometric system is approx. 0.01 μm. The surface profiles were recorded over a scan length of 48 or 15 mm. A cut off wavelength of 8 mm for a scan length of 48 mm was used to separate between roughness and waviness profile. The evaluation of the mechanical profile measurement according to typical roughness parameters – e.g. average roughness Ra – gives an integrated information about the surface structure. In comparison to roughness parameters, Fourier techniques (FFT) yield a more detailed characterization of the surface structure.

For wavelengths from 10 to 1 mm (integral 1, long waviness) and from 1 to 0.1 mm (integral 2, short waviness), the intensities of the autopower spectra are added up and used for further evaluation of the surface structures (see e.g. [14]).

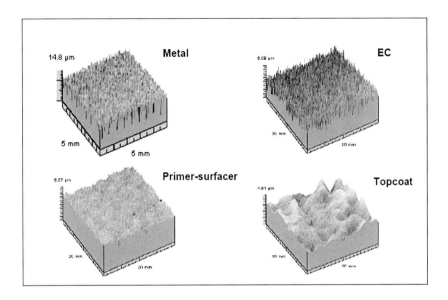

Fig. 1: Topographies of metal substrate and coating layers

In fig. 1 typical surface topographies from substrate to topcoat are shown in a pseudo-3-dimensional presentation. The sample is moved by a precise positioning table for a small distance between two line scans. In general, a decrease of the amplitudes and a change in wavelengths can be observed.

2.2 Optical surface characterization

2.2.1 wave-scan

The optical determination of coating structures was carried out by the wave-scan *dual* or with the former instrument wave-scan DOI. Here, the measuring principle is based on the modulation of the light of a small laser diode reflected by the surface structures of the sample. The laser light shines on the surface at an angle of 60°, and the reflected light is detected at the gloss angle (60° opposite). During the measurement, the 'wave-scan' is moved across the sample surface over a scan length of approx. 10 cm. The signal is divided into 5 wavelength ranges in the range of 0.1 to 30 mm and processed by mathematical filtering. For each of the 5 ranges a characteristic value (Wa 0.1-0.3 mm, Wb 0.3-1.0 mm, Wc 1.0-3.0 mm, Wd 3.0-10 mm, We 10-30 mm) as well as the typical wave-scan-values longwave (LW, approx. 1-10 mm) and shortwave (SW, approx. 0.3-1 mm) is calculated. Low wave-scan-values mean a smooth surface structure. Additionally a LED light source is installed and illuminates the surface under 20° after passing an aperture. The scattered light is detected and a so-called dullness value (du, < 0.1 mm) is measured. A change to an IR-SLED for the low gloss area allows to measure samples with medium glossy surfaces.

3 Metal substrates

In this part an example of former studies on steel substrates [10] is presented. Three different phosphated steel surfaces with different roughness levels were evaluated. The samples were chosen in order to cover the range of roughness from smooth to rough substrates (Ra 0.8 to 2.3 µm) supplied for automotive body sheets, where – typically – sheets with an average roughness between Ra = 0.8 µm (smooth) and Ra = 1.6 µm (middle, upper limit) were used. All samples were coated with cathodic electrodeposition paint, primer-surfacer and automotive topcoat with typical layer thicknesses. One set of painted substrates was baked in horizontal position, a second set vertically.

As shown in fig. 2, the integrated values increase as a function of substrate surface roughness and baking position. For the vertical baking position, a strong enhancement of the long waves can be observed, while the short wave values hardly increase compared to the samples baked horizontally. The short wave values depend on substrate conditions (roughness), not strongly on baking position.

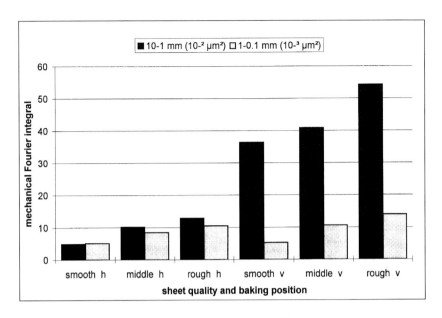

Fig. 2: Mechanical Fourier integrals obtained by measurements on topcoats

4 Investigations on plastics coatings

Investigations on plastics coatings are of special interest to achieve a similar structure development on a painted car not depending on the different substrate materials (steel or plastics) used. Against this background parameters influencing the surface of plastics parts are becoming recently more important.

4.1 Measurements on reinforced plastics (structure effect)

Polymeric substrates with different amount of glass fibers were painted with a typical coating system for plastics. The complete coatings were measured by wave-scan and mechanical profilometry [12]. Optical and mechanical measurements (Ra) show both the same trend, an increasing measuring value of structure for an increasing amount of glass fiber. In addition to the effect of the amount of fiber content also a dependence on the type of the used fiber can be observed in a different study [13].

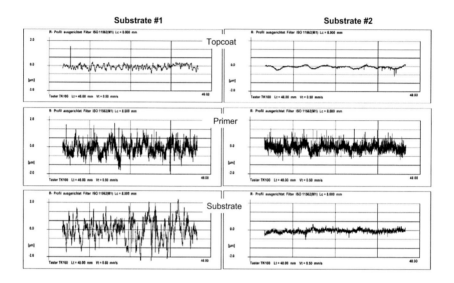

Fig. 3: Comparison of two substrates and coating layers (mech. profilometry, roughness profiles, scan length 48 mm, y-scale +- 2 μm)

This effect is also demonstrated in fig. 3 showing a comparison of two substrates with high and low surface structure. The amplitude of the structures is reduced, but the basic structure of the substrate is partially transferred through each single coating layer and influences finally the topcoat appearance.These results are very important due to the effect, that the substrate structure clearly influences the final coating structure. Based on surface structure measurements a pre-selection of plastics substrates can be done to obtain an optimal structure (smoothness) of the coating system applied [15].

5 Investigations on medium glossy surfaces

Several test sets have been measured to investigate the effects of coating variation (EC and primer-surfacer) and substrate topographies on the different coating layers. These studies are described in detail in [18]. The investigations and main results for varying the steel substrate will be summerized in this chapter.

5.1 Variation of substrate

Most of the 28 studied substrates were in the range which is regularly used in the automotive industry for outer car body parts (Ra 0.9-1.5 μm, RPc > 60). In fig. 4 the profilograms of substrates and the corresponding EC (electro coating) surfaces are demonstrated by 4 selected samples.

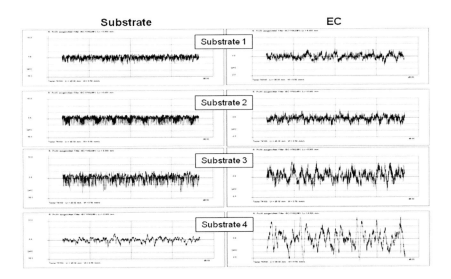

Fig. 4: Roughness profile of panel and EC surfaces (variation substrate) Scan length 48 mm, y-scale +- 10 μm (substrate) +- 2 μm (EC)

The Ra values of the substrates 1-3 were in a roughness range from 0.9-1.5 μm, with peak counts of 60-75 points/cm. Substrate 4 shows a good Ra value of 0.8 μm, but a very low number of 12 peaks/cm. The painting of this substrate is more difficult because of the low peak number and would not be used for the outer shell of a car body. The EC application as well as the whole following coating process was carried out with the same material and under identical application conditions.

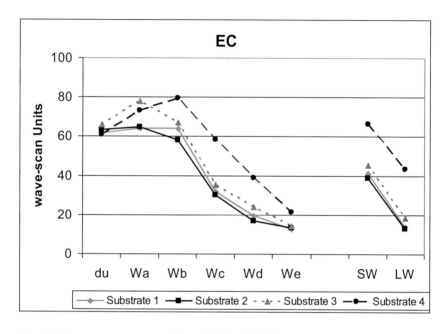

Fig. 5: Structure spectra, SW and LW of EC (variation substrate)

The measurement of the EC (fig. 5) with the wave-scan dual could be carried out without any difficulty and shows for the 4 example substrates comparable ranking of EC and topcoat structure (fig. 6).

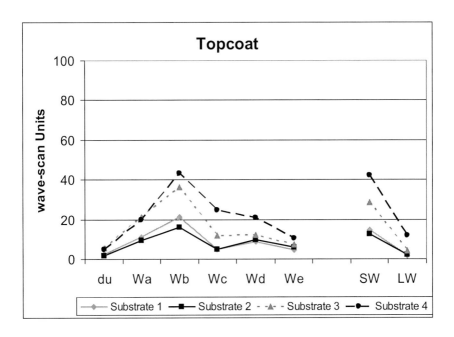

Fig. 6: Structure spectra, SW and LW of topcoats (variation substrate), horizontal application

Considering the whole sample set, the linear correlation coefficient of profilometric structure measurement between substrate and EC structure is r≈0.9 for the long wave range (integral 1) and r≈0.7 for the short wave range (integral 2). There is a correlation coefficient of 0.8 between EC and clearcoat structure for horizontal application [18].

6 Further studies and summary

Further studies considering the influence of sheet-type, substrate roughness, deformation and paint system on the appearance of the painted surface were summarized in [15]. Additional investigations by variation of paint systems with lower gloss (EC, primer-surfacer) can be found in [18].

Influences of the forming/deep-drawing process yielding to an increased waviness of metal substrates were reported in [19].

Investigations of the mechanisms for the formation of surface structures on paint layers are considering for example the flow behaviour of the liquid paint driven by viscosity, surface tension and other parameters. Modeling and simulations of the interrelations of intrinsinc physical and process parameters are used to understand the causes of paint film appearance [20-22].

The mechanical profilometry can be applied for all substrates and coating layers. The use of the wave-scan dual is limited by dullness and a distinct surface structure. The optical investigation of EC is possible, if the surface structure of the used substrate in combination with the self structure of the coating material is not too distinguished. In the daily work primer-surfacers can usually be investigated without reservation.

References

01.   D. W. Boyd, Proc. XIII. Int. Conf. Org. Coat. Sci. Techn., Athens, (1987) 59

02.   F. Fister, N. Dingerdissen, C. Hartmann, Proc. XIII. Int. Conf. Org. Coat. Sci. Techn., Athens, (1987) 113

03.   K. Armbruster, M. Breucker, Farbe Lack, 95 (1989) 896

04.   T. Nakajima, Y. Yoshida, Y. Miyoshi, T. Azami, Proc. XVII. Int. Conf. Org. Coat. Sci. Techn., Athens (1991) 227

05.   W. Geier, M. Osterhold, J. Timm, Metalloberfläche, 47 (1993) 30

06.   A. F. Bastawros, J. G. Speer, G. Zerafa, R. P. Krupitzer, SAE Technical Paper Series 930032

07. J. Timm, K. Armbruster, M. Osterhold, W. Hotz,
Bänder, Bleche, Rohre, 35 (1994), No. 9, 110

08. M. Osterhold, W. Hotz, J. Timm, K. Armbruster,
Bänder, Bleche, Rohre, 35 (1994), No. 10, 44

09. M. Osterhold, Prog. Org. Coat., 27 (1996) 195

10. M. Osterhold, Mat.-wiss. u. Werkstofftech., 29 (1999) 131

11. O. Deutscher, K. Armbruster, Proc. 3rd Stahl-Symposium,
Düsseldorf, Germany (2003)

12. H. Stegen, M. Buhk, K. Armbruster, Paper TAW Seminar
„Kunststofflackierung – Schwerpunkt Automobilindustrie",
Wuppertal, Germany (2002)

13. K. Armbruster, H. Stegen, Proc. DFO Congress
„Kunststofflackierung", Aachen, Germany (2004) 78

14. M. Osterhold, K. Armbruster, Proc. DFO Congress
„Qualitätstage 2005", Berlin, Germany (2005) 4

15. M. Osterhold, K. Armbruster, Prog. Org. Coat., 57 (2006) 165

16. O. Deutscher, BFI, Düsseldorf, Carsteel-Bericht (2008)

17. M. Osterhold, K. Armbruster, Proc. DFO Congress
„Qualitätstage 2008", Fürth, Germany (2008) 7

18. M. Osterhold, K. Armbruster, Prog. Org. Coat., 65 (2009) 440

19. D. Weissberg, Proc. DFO Congress
„21. Automobil-Tagung 2014", Augsburg, Germany (2014)

20. C. Hager, M. Schneider, U. Strohbeck, Proc. ETCC 2012,
Lausanne, Switzerland (2012)

21. M. Hilt, M. Schneider, „Die Entstehung von Lackfilmstrukturen verstehen", www.besserlackieren.de, 07.03.2014

22. O. Tiedje, Proc. DFO Congress „21. Automobil-Tagung 2014", Augsburg, Germany (2014)

Improving Rheology Measurement
Yield Point and Thixotropy

Abstract

Yield point and thixotropy influence important materials properties. The Working Group "Rheology" of the Standards Committee coatings and coating materials (NAB) within the DIN (German Institute of Standardization e. V.) has followed up intensively on the measuring characterization of yield points and thixotropy in the last years. Two technical reports on these items have been prepared. Procedure and some results will briefly be summerized in this paper.

1 Introduction

Many applicational and technological properties are influenced by the flow behaviour of the coating. High product quality can only be guaranteed through an exact knowledge of the rheological behaviour of the coating and the used raw materials, respectively. In view of the increasing use of waterborne systems flow anomalies as thixotropy, yield points or also viscoelastic behaviour can be observed more often. Such behaviour is not normally observed in conventional, solvent-borne coatings. If, however, so-called SCA agents (Sagging Control Agents) are added to directly control rheological properties, phenomena like thixotropy, yield points or viscoelasticity can appear as well [1- 3].

Yield point and thixotropy influence important materials properties as storage stability, pumping behaviour or levelling and flowing. Against this background the Working Group "Rheology" of the Standards Committee coatings and coating materials (NAB) within the DIN (German Institute of Standardization e. V.) has followed up intensively on the measuring characterization of yield points and thixotropy in the last years. Two technical reports on these items have been prepared [3, 8].

The measuring possibilities of characterizing the rheological properties with rotational rheometers concerning yield point and thixotropy will be presented in this paper.

2 Definition and importance of yield point and thixotropy

The yield point is defined as the lowest shear stress above which the behaviour of a material, in rheological respect, is like that of a liquid; below the yield point its behaviour is like that of an elastic or viscoelastic body.

Thixotropy describes a flow behaviour, where the rheological parameters (viscosity) decrease due to a mechanical constant load to a timely constant limiting value and after reducing the load, the initial state is completely reached depending on time. In practice, only a limited time frame is considered in which the initial state is not always reached.

With yield point and thixotropy important material properties can be characterized, e.g.

- Effectiveness of rheological additives
- Storage stability (e.g. against sedimentation, separation, flocculation)
- Behaviour when starting to pump
- Wet film thickness
- Levelling and flowing behaviour
  (e.g. without brushmarks and sag formation)
- Orientation of effect pigments

3 Methods for determining the yield point

The individual methods for determining the yield point are summerized and critically discussed in the DIN technical report 143 [4] . The presented results for evaluating yield points in this report base on interlaboratory tests, which were carried out by the participants of the Working Group "Rheology" of the Standards Committees "Pigments and Extenders" and "Coatings and Coatings Materials" at DIN.

In first preliminary tests different waterborne basecoats with low and dispersions with distinctly higher yield points have been examined as well. It was found that some methods showed unexpectedly good qualitative relationships. On the other hand, some participants reported problems with the preparation of the samples. In addition, in the course of the preliminary tests certain methods of measurement have been found to be unsuitable for the samples examined and were therefore no longer considered. In this connection, the method of maximum viscosity and the method using a vane measuring system have to be mentioned. Also the method for determining the yield point using a linear stress ramp was not helpful as there are not enough measuring points in the lower measuring range. Also evaluation procedures based on traditional regression methods (e.g. according to Bingham or Herschel-Bulkley) were not considered in further tests. The results depend too strongly on the theoretical model used and the measuring specifications (ramps) (according to [3, 4]).

3.1 Tangent method in a representation of a lg γ / lg τ diagram

Based on the experiences made in preleminary tests, the participants agreed on the method „stress ramp" to be used in a continued interlaboraty test. The results are presented in the technical report 143. Therefore, below and above the assumed yield point one decade for the evaluation should be available. The (logarithmic) shear stress ramp should begin at least one decade below the assumed yield point and should reach at least one decade beyond the yield point value.

Exact test conditions have been agreed by the Working Group and definitely been specified for all participants of the comparative testing programme (see [4]). Five different samples have been examined in total: two waterborne basecoats with low yield points of a magnitude of 1 Pa or smaller, two dispersions and one sample with well known yield point provided by the Physikalisch-Technischen Bundesanstalt (PTB, The National Metrology Institute of Germany).

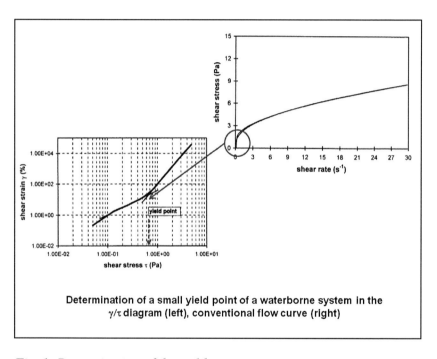

Determination of a small yield point of a waterborne system in the γ/τ diagram (left), conventional flow curve (right)

Fig. 1: Determination of the yield point

If a yield point is existing, a straight line can be observed in the range of low shear stress, shear stress τ and shear deformation/strain γ are then proportional at low values. The investigated material shows consequently a reversible linear-elastic deformation behaviour (Hooke's elasticity law). At higher stresses the structure at rest breaks down, the deformation then becomes disproportionately high, and the material shows irreversible viscoelastic or viscous flow behaviour [5-7]. The yield point has exceed if the measuring points are no longer on a straight line ("tangent"). If it is possible to set up a second tangent through the measuring points at high deformations – also in the flow range – , then the crossover point of the tangents is evaluated as yield point (fig. 1) [3]. This evaluation method is described in [4] in more detail.

In summary, good results have been obtained which was due to the specfed time schedule of the tests and the previously specified measuring conditions.

For a clear characterization of yield points for various products, detailed test specifications for different substance classes have to be developed.

4 Measuring characterization of thixotropy

After finishing the investigations and the preparation of the technical report on yield point determination, 2005 first tests concerning "Thixotropy" were initiated by the DIN Working Group "Rheology". At the beginning, the target was to prove the fundamental suitability, in a second comparative testing programme (interlaboratory tests) in 2009/2010 spccial emphasis was given to the determination of precision data for different measuring procedures. For that purpose a waterborne basecoat and a clearcoat were investigated with four different measuring methods. A Newtonian liquid of the PTB was used as a reference. The final technical report on thixotropy was published in September 2012 [8].

A widely spread method for determining the thixotropy bases on the registration of flow curves with defined parameters for the measuring procedure. To fix are the time in which a certain maximal shear rate has to be reached (ramp time for up-curve, number of measuring points etc.), holding time at max. shear rate and time for the down-curve. Furthermore, it is to decide, wether a linear or logarithmic ramp has to be used (continuously or step-like). Besides a precise temperature control, a waiting time or pre-shearing just before the actual measurement could be important. This procedure by means of flow curves is called hysteresis method or thixotropic loop. Here, the area between up- and down-curve is evaluated as a measure for the grade of thixotropy.

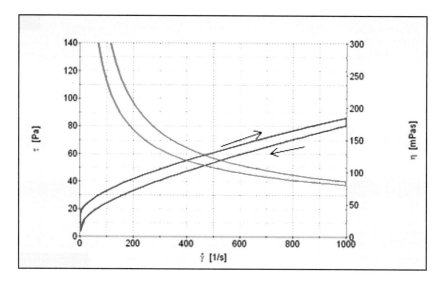

Fig. 2: Flow curve (red) and viscosity curve (green) of a blue metallic waterborne basecoat

In fig. 2 the flow curve (red) of a blue waterborne basecoat is shown as a typical example of a thixotropic coating system. The added arrows indicate each the direction of increasing (up-curve →) and decreasing (down-curve ←) shear rate, respectively. A marked hysteresis as an indicator of thixotropy can be clearly detected. In addition, two marked yield points can be observed in the low shear rate range. Also shown is the respective (apparent) viscosity run (green curves). The viscosity η decreases for higher shear rates $\dot{\gamma}$ to a value of approx. 80 mPas at

1000 s^{-1} , and after reducing the load, i. e. at low shear rates, to increase again.

Fig. 3 shows the comparison of measurements of a waterborne basecoat with linear and logarithmic ramp [8]. Marked differences can be observed for the hysteresis areas.

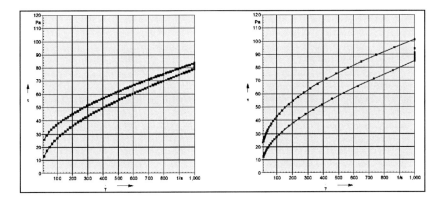

Fig. 3: Flow curves of a waterborne basecoat (left linear, right logarithmic ramp)

Due to the increasing development of controlled-stress rotational rheometers in the last years, combinations of shear and/or oscillation procedures are applied. Such experiments are usually divided into three segments (see also fig. 4 [8]). In this figure, the preset profile and the measuring results are schematically shown for the rotational case. In a first step the sample under investigation is subjected to a low shear rate (rotation), oscillation or shear stress, followed by a severe loading under rotation with high shear rate, and finally the recovery phase at a low load under rotation (shear rate), oscillation or shear stress (recovery/structure re-build).

In the second technical report of the DIN working panel several items concerning the determination of thixotropy are discussed in detail [8].

Rheological methods and the results of comparative testings of up to nine different laboratories are presented.

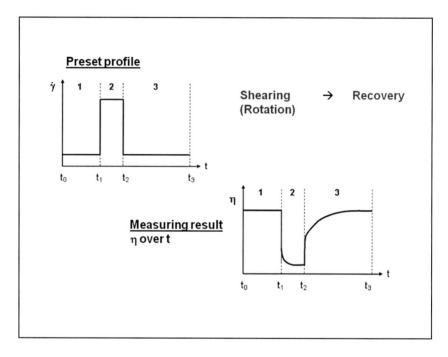

Fig. 4: Schematical preset profile and measuring result for rotation: time-dependent viscosity function of a thixotropic substance, (1) at low, (2) at high shear load with reduction of structure, and (3) again at low shear load with recovery of structure

The results show, that the method for the determination of thixotropy using flow curves (hysteresis area) is only partly suitable. Methods combining low and severe shear loads followed by the structure recovery phase at a low load allow a better evidence concerning thixotropy.

Considering values in the phase of structure recovery (3rd segment) related to values in the 2nd segment (high shearing), suitable characteristic values like TI (thixotropy index) or SRI (structure recovery index) can be determined. Precise evaluation methods and proposals for carrying out the measurements as well as presets for the measuring parameters are descript in detail in [8].

References

01. M. Osterhold, Prog. Org. Coat., 40 (2000) 131

02. M. Osterhold, Farbe Lack, 116 (2010), No. 9, 33

03. M. Osterhold, Proc. DFO „Qualitätstage", Köln, Germany (2011) 99

04. H. Bauer, E. Fischle, L. Gehm, W. Marquardt, T. Mezger and M. Osterhold, DIN-Fachbericht 143 – Moderne rheologische Prüfverfahren – Teil 1: Bestimmung der Fließgrenze, Grundlagen und Ringversuch, Beuth-Verlag, Berlin (2005)*; and summary of the report

05. L. Gehm, Rheologie, Vincentz, Hannover (1998)

06. G. Schramm, Einführung in Rheologie und Rheometrie, Haake, Karlsruhe (1995)

07. T. Mezger, Das Rheologie-Handbuch, Vincentz, Hannover (2000)

08. E. Fischle, E. Frigge, L. Gehm, H. Klee-Wohlenberg, C. Küchenmeister, T. Mezger, M. Osterhold, T. Remmler, U. Weckenmann, H. Wolf and R. Worlitsch, DIN SPEC 91143-2 – Moderne rheologische Prüfverfahren – Teil 2: Thixotropie, Bestimmung der zeitabhängigen Strukturänderung – Grundlagen und Ringversuch, Beuth-Verlag, Berlin (2012)*

*Text in German and English

Methoden zur Charakterisierung der Kratzbeständigkeit

Abstract

Die Kratzbeständigkeit ist eine der wichtigsten Eigenschaften, um die mechanische Qualität eines Lacksystems zu charakterisieren. In diesem Artikel werden die relevante Messtechniken kurz behandelt mit Anwendung in der Automobil- und Lack-Industrie. Techniken, die auf Einzel-Kratz-Tests zur Bestimmung der Kratzbeständigkeit beruhen, werden mit mehr einfachen Methoden, wie dem Waschbürsten-Test (Amtec), verglichen.

1 Einleitung

Zur Erzielung einer hohen optischen Qualität von Klarlackoberflächen werden neben einem guten Verlauf und hohem Glanz insbesondere Beständigkeiten gegen mechanische Beanspruchungen – Steinschlag und Verkratzungen – gefordert. So erzeugen z. B. in einer Automobilwaschstraße Waschbürste und Schmutz Kratzer mit einer Breite von wenigen µm bei einer Tiefe bis zu einigen Hundert nm.

Vor diesem Hintergrund sind verschiedene Messmethoden in den letzten 15 bis 20 Jahren in der Automobil- und Lack-Industrie diskutiert worden. Ziel war es, eine klare Charakterisierung der Kratzbeständigkeit von Klarlacken zu erhalten. Verfahren, die einen einzelnen Kratzer erzeugen sind zum Ende der 1990er Jahre entwickelt und verbessert worden (Micro- oder Nano-Scratch-Methode) [1-8]. Dem gegenüber stehen mehr praxisorientierte Verfahren, die auf relativ einfachen Methoden beruhen bzw. versuchen, nah an der Realität (z. B. Waschbürsten-Verfahren) zu prüfen.

2 Charakterisierungsmethoden

2.1 Laborwaschanlage (Amtec)

Der zu prüfende Klarlack (lackiert auf Standard-Blechen) wird hierbei 10x unter einer rotierenden Autowaschbürste hin- und herbewegt. Während des Waschvorgangs wird die Bürste (PE) mit Waschwasser besprüht. Da die Prüfbleche im Labor frei von Verschmutzungen sind, wird dem Waschwasser als Ersatz für den Straßenschmutz in definierter Menge Quarzmehl beigefügt (Amtec Methode). Zur Beurteilung der Kratzfestigkeit wird die Glanzmessung in der 20°-Geometrie herangezogen. Dabei wird der Anfangsglanz und nach Reinigung der Endglanz gemessen. Der auf den Anfangsglanz bezogene prozentuale Restglanz ist ein Maß für die Kratzfestigkeit. Hohe Werte entsprechen hierbei einer guten Resistenz, das Gerät ist schematisch in Abb. 1 dargestellt [8]. Eine kommerzielle Laborwaschanlage wird von Amtec-Kistler (Deutschland) hergestellt.

Diese Methode wurde in einer Projekt-Arbeitsgruppe der DFO (Deutsche Forschungsgesellschaft für Oberflächenbehandlung) vor zwei Jahrzehnten entwickelt und ist in Standard DIN EN ISO 20566 spezifiziert worden. Mehr Details können beispielsweise in [9] gefunden werden.

2.2 Crockmeter

Das Crockmeter wird bei großen Autoherstellern angewendet und stellt eine andere Form der Belastung dar. Das Crockmeter ist ein Standardgerät der American Association of Textile Chemists and Colorists

44

(AATCC). Es dient in erster Linie zur Prüfung der Farbechtheit und Reibfestigkeit von Textilien.

Das Gerät (Atlas Materials Testing) ist mit einem Elektromotor ausgestattet, wodurch eine gleichmäßige Belastung von 60 Doppel-Hüben pro Minute erreicht wird. Die Probe wird auf einem flachen Sockel befestigt. Es erfolgt eine lineare Reibbelastung mit einem Zylinder (für andere Geometrien siehe DIN 55654) und speziellen Testmaterialien. Der Durchmesser des Zylinders beträgt 16 mm, die Auflagekraft 9 N. Es werden 10 Doppelhübe auf einer Reiblänge von 100 mm durchgeführt. Auch bei dieser Methode wird der prozentuale Restglanz als Maß für die Verkratzung herangezogen.

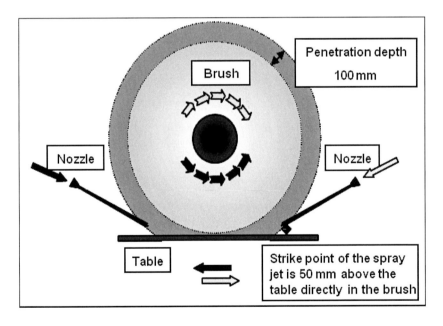

Abb. 1: Laborwaschanlage (Amtec)

In [8] kann eine gute Korrelation beider Messverfahren, Crockmeter und Laborwaschanlage, im Bereich guter und schlechter Kratzfestigkeit beobachtet werden. Im Bereich mittlerer Werte allerdings ist kein eindeutiger Zusammenhang feststellbar. Ursache hierfür könnte der unterschiedliche Beanspruchungsgrad bei der Verkratzungsprüfung sein. So stellt die Amtec-Prüfung eher die stärkere Beanspruchung dar.

2.3 Rota-Hub-Methode

Der Rota-Hub-Scratch-Tester wurde von der Fa. Bayer AG vor etwa 15 Jahren entwickelt. Hierbei bewegt sich ein Schlitten in x- und y-Richtung. An dem Schlitten befindet sich zusätzlich eine rotierende Scheibe mit dem Verkratzungsmaterial (z. B. Papier), die auf die Probenplatte abgesenkt wird. Die Probenoberfläche wird somit durch eine rotierende Scheibe, die sich zusätzlich in x- und y-Richtung bewegt, belastet. Sowohl die Vorschub- und Umdrehungsgeschwindigkeiten als auch die Anpresskraft der rotierenden Scheibe sind frei wählbar und wurden optimiert.

Als Reibmaterial zeigte sich Kopierpapier für die Automobilindustrie als geeignetes Verkratzungsmaterial, da hiermit waschstraßenähnliche Verkratzungen erhalten werden. Als Schädigung erhält man ein mäandrisches Kratzmuster [10].

2.4 Micro-Scratch-Experimente

Bei Micro-Scratch-Experimenten können einzelne Kratzer mit charakteristischem und praxisgerechtem Erscheinungsbild erzeugt werden. Die Eindringtiefen sind abhängig von der verwendeten Kraft und dem Eindringkörper (Indentor). Sie liegen üblicherweise in einem Bereich um 1 μm und kleiner. Zum Anbringen der Verkratzung wird ein Diamantindentor (Spitzenradius 1 bis 2 μm) mit definierter Normalkraft auf die Probenoberfläche gedrückt, während die Probe mit konstanter Geschwindigkeit linear unter dem Indentor bewegt wird (Abb. 2).

Grundsätzlich können während des Verkratzens die Tangentialkräfte und die Eindringtiefen vermessen und mikroskopisch oder zusätzlich per Videokamera die Deformationen oder Beschädigungen beobachtet werden. Mittels AFM-Technik können die entstandenen Kratzspuren im Profil ausgemessen werden. Zusammen mit den gemessenen Parametern während des Kratzens lassen sich physikalische Kenndaten berechnen, aus denen Rückschlüsse über das elastische und plastische Verformungsverhalten sowie das Bruchverhalten gezogen werden können.

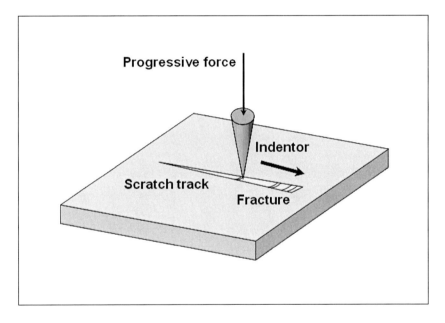

Abb. 2: Prinzip von Single-Scratch-Tests

2.4.1 Tests mit ansteigender Kraft

In diesem Zusammenhang wurde im DuPont Marshall Lab. (Philadelphia) eine Methode entwickelt, die einen einzelnen Kratzer auf einer Oberfläche erzeugt und auswertet (Nano-Scratch-Verfahren). Dieses

Verfahren ist bei entwicklungsbegleitenden Untersuchungen von Klarlacken eingesetzt worden.

Im Rahmen eines Projektes beim Forschungsinstitut für Pigmente und Lacke e.V. (FPL, nun Teil von IPA/Stuttgart) wurde ein kommerzielles Gerät der Schweizer Firma CSM (nun Teil von Anton Paar) um das Jahr 2000 und später getestet. Dieses Gerät basiert auf dem Verfahren, das im DuPont Marshall Lab. entwickelt wurde. Ziel verschiedener Projekte – in dem Automobil-, Lack- und Rohstoffhersteller mitarbeiteten – war die Bewertung des Nano-Scratch-Testers bezüglich Reproduzierbarkeit, Genauigkeit und Aussagekraft der Messergebnisse im Hinblick auf eine praxisgerechte Ermittlung der Kratzbeständigkeit

Zentraler Punkt bei diesem Verfahren ist die Ermittlung der sogenannten kritischen Kraft (Critical load), bei der erste irreversible Risse oder Brüche (Cracks, Fracture) im Material entstehen und die damit den Übergang von plastischer Verformung zu deutlichen/bleibenden Verletzungen kennzeichnet. Hierzu wird die Normalkraft stetig erhöht mit gleichzeitiger Detektion von Eindringtiefe und Tangentialkraft. Der Übergang von plastischer Verformung zum Fracture-Bereich zeigt sich z. B. durch Unstetigkeiten und Fluktuationen im detektierten Kräfteverlauf und bei der Eindringtiefe. Durch zusätzliche optische bzw. AFM-Untersuchungen kann ebenfalls dieser Übergangsbereich ausgewertet werden. Neben der Bestimmung der kritischen Kraft wird als typische Maßzahl die verbleibende Eindringtiefe nach Verkratzung – typischerweise bei einer Normalkraft von 5 mN – im Bereich der plastischen Verformung bestimmt.

In Abb. 3 ist der Übergangsbereich eines Acrylat/Melamin-Systems dargestellt. Die obere Abbildung zeigt die AFM-Aufnahme der eigentlichen Kratzfurche, in der deutlich Risse (Cracks) zu erkennen sind. Die untere Aufnahme wurde mit dem Lichtmikroskop in etwa 1000-facher Vergrößerung erstellt, siehe auch [8, 11, 12].

Werden Werte aus den Nano-Scratch-Experimenten mit Ergebnissen aus Waschstraßen-Untersuchungen verglichen (s. Abb. 4 [8]), zeigt sich ein gewisser Zusammenhang mit der kritischen Kraft. Daraus lässt sich ableiten, dass die Amtec-Prüfung eher eine deutliche Verletzung der Oberfläche verursacht als im plastischen Bereich zu beanspruchen [13].

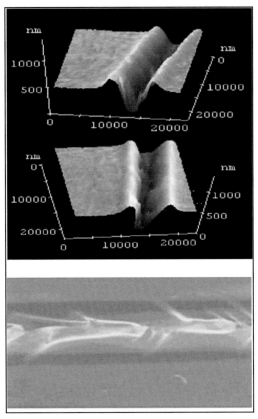

Abb. 3: AFM (oben) und mikroskopische (Lichtmikroskop) Bilder (unten) von Cracks in einem Acrylat/Melamin System

3 Dynamisch-Mechanische Analyse

Auf der Zeitskala kann bei bestimmten Klarlacksystemen ein teilweises Ausheilen der Verkratzung beobachtet werden, was in der Literatur als

Reflow-Effekt bezeichnet wird [14]. Zur physikalischen Deutung dieses Effektes können thermische Relaxationsphänomene herangezogen werden. Im Zusammenhang mit Kratzbeständigkeit spielt die Vernetzungsdichte der Klarlacke eine entscheidende Rolle. Als Methode zur Bestimmung der Vernetzungsdichte hat sich die Dynamisch-Mechanische Analyse DMA fest etabliert [14-16].

Anhand des Verlaufes des Speichermoduls E´ als Funktion der Temperatur können hierbei verschiedene Bereiche charakterisiert werden. So schließt sich dem Glasübergangsbereich der je nach Vernetzungsgrad mehr oder weniger ausgeprägte gummielastische Bereich an. Der Wert des Speichermoduls in diesem Bereich ist mit der Vernetzungsdichte verknüpft. Als quantitatives Maß wird hierfür oft der Wert von E´ im lokalen Minimum gewählt.

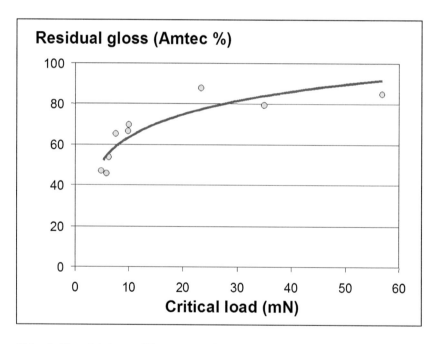

Abb. 4: Vergleich von Werten aus Amtec-Messungen mit Werten der kritischen Kraft aus Nano-Scratch-Messungen

Die Untersuchungen der Klarlacke erfolgten im Zugversuch an freien Filmen mit einer DMA 7 (Perkin Elmer). In diesem Zusammenhang durchgeführte Messungen an Lacksystemen sind z. B. in [16] beschrieben.

In Abb. 5 sind die Ergebnisse der DMA-Untersuchungen den jeweiligen Werten aus der Waschstraßen-Simulation gegenübergestellt [8].

Zu beobachten ist hierbei ein deutlicher Zusammenhang zwischen E' als Maß für die Vernetzungsdichte und dem Wert der Verkratzung. So steigt mit zunehmender Vernetzungsdichte die Güte des Verkratzungsniveaus (hoher Restglanz) [17].

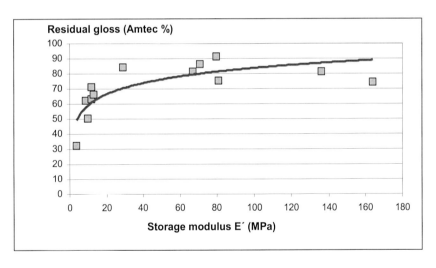

Abb. 5: Zusammenhang von Kratzfestigkeit (Amtec-Werte) mit Vernetzungsdichte (Speichermodul E')

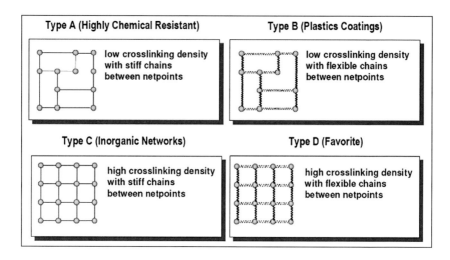

Abb. 6: Unterschiedliche Netzwerktypen

Abb. 6 zeigt einen kurzen Vergleich unterschiedlicher Netzwerk-Typen und zusammengefasste Anmerkungen über deren Eigenschaften [18, 19, 20]; mit Typ D als Favorit für Automobil-Klarlacke.

4 Einfluss der Bewitterung

Der Einfluss der Bewitterung auf die Kratzbeständigkeit wird in Abb. 7 demonstriert, dargestellt ist die Glanzerhaltung nach Amtec-Prüfung für verschiedene Klarlacke als Funktion der Expositionszeit bei SAEJ1960-Bewitterung. Alle untersuchten Proben zeigen eine signifikante Abnahme des Restglanzes nach kurzer Expositionszeit. Eine klare Unterscheidung zwischen den Klarlacken mit (CC1, CC2, CC3, CC1c) und ohne Lichtschutzmittel (CC1a, CC1b, CC2a) kann beobachtet werden. Anschließend fällt der Restglanz leicht ab [21, 22]. Andere Studien können z. B. in [23] gefunden werden.

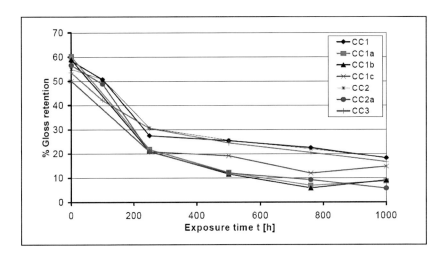

Abb. 7: Glanz-Erhaltung (Amtec) nach SAEJ1960-Bewitterung

5 Zusammenfassung

Verschiedene Methoden zur Bestimmung der Kratzbeständigkeit von Klarlacken wurden in diesem Artikel kurz vorgestellt. Methoden, die einen einzelnen Kratzer erzeugen (Micro- oder Nano-Scratch-Verfahren), wurden in ihrer Aussagekraft mehr praxisorientierten Verfahren gegenübergestellt, die im Vergleich auf relativ einfachen Methoden beruhen (z. B. Amtec-Waschbürsten-Verfahren). Die Methoden wurden zusammengefasst betrachtet, auch unter Berücksichtigung anderer physikalischer Eigenschaften wie Vernetzungsdichte oder Bewitterung.

Untersuchungen der letzten Zeit zeigen, dass Kratzbeständigkeit immer ein wichtiger Faktor zur Charakterisierung physikalischer Eigenschaften von Lackoberflächen ist, z. B. bei der Analyse von Automobil-Klarlacken, die Silan-modifizierte blockierte Isocyanate enthalten [24]. Ihre Ergebnisse zeigen, dass eine enge Korrelation existiert zwischen den Daten der Kratzbeständigkeit aus Waschstraßen-Prüfung und Nano-Scratch-Tests für bestimmte Klarlacke.

Auch die Kratzbeständigkeit von Aussen-Klarlacken und Polycarbonat-Hardcoats wurde in jüngerer Zeit untersucht und diskutiert in [25].

In der Literatur wird oft nicht klar unterschieden zwischen den Ausdrücken *Mar* und *Scratch* Resistance. *Mar* sollte für leichte Oberflächenbeschädigungen benutzt werden, die im realen Umfeld auftreten und üblicherweise klein sind, während *Scratch* für mittlere oder schwere Verletzungen gewählt werden sollte [26]. Wie auch immer, verschiedene Autoren umgehen diese Unsicherheit, indem sie *Scratch/Mar* schreiben.

Literatur

01. G. S. Blackmann, L. Lin, R. R. Matheson, ACS Symposium Series: Microstructure and Tribology of Polymer Surfaces (1999)

02. L. Lin, G. S. Blackman, R. R. Matheson, Prog. Org. Coat., 40 (2000) 85

03. K. Adamsons, G. Blackmann, B. Gregorovich, L. Lin, R. Matheson, Prog. Org. Coat., 34 (1998) 64

04. J. L. Courter, Proc. 5th Nürnberg Congress, Nuremberg, Germany, (1999) 351

05. E. Klinke, C. D. Eisenbach, Proc. 6th Nürnberg Congress, Nuremberg, Germany, (2001) 249

06. L. Lin, G. S. Blackman, R. R. Matheson, Materials Science and Engineering, A317 (2001) 163

07. G. Wagner, M. Osterhold, Mat.-wiss. u. Werkstofftech., 30 (1999) 617

08. M. Osterhold, G. Wagner, Prog. Org. Coat., 45 (2002) 365

09.   E. Fischle, Farbe Lack, (2009), No. 04

10.   E. Klinke, M. Kordisch, C.D. Eisenbach, T. Klimmasch,
        Farbe Lack, 108 (2002), No. 2, 55

11.   M. Osterhold, Europ. Coat. Journal, (2005), No. 09, 34

12.   M. Osterhold, Progr. Colloid Polym. Sci., 132 (2006) 41

13.   K.-F. Dössel, SURCAR, Cannes, France, 2001

14.   R. Gräwe, W. Schlesing, M. Osterhold, C. Flosbach,
        H.-J. Adler, Europ. Coat. Journal, (1999), No. 3, 80

15.   W. Schlesing, Farbe Lack, 99 (1993) 918

16.   W. Schlesing, M. Buhk, M. Osterhold, Prog. Org. Coat.,
        49 (2004) 197

17.   E. Frigge, Farbe Lack, 106 (2000), No. 7, 78

18.   M. Osterhold, Vortragstagung der GDCh-Fachgruppe Anstrich-
        stoffe und Pigmente, Eisenach, Germany, 2005

19.   C. Flosbach, XXVIII Fatipec Congress, Paper II.C-1, Budapest,
        Hungary (2006)

20.   K.-F. Dössel, in: Automotive Paints and Coatings, 2nd Ed.,
        H.-J. Streitberger, K.-F. Dössel (Eds.), Wiley-VCH (2008)

21.   M. Osterhold, B. Bannert, W. Schubert, T. Brock,
        Macromol. Symp. 187, (2002) 823

22.   B. Bannert, M. Osterhold, W. Schubert, T. Brock,
        Europ. Coat. Journal, (2001), No. 11, 30

23.   C. Seubert, M. Nichols, K. Henderson, M. Mechtes,
      T. Klimmasch, T. Pohl, J. Coat. Technol. Res., 7 (2010), 159

24.   S. M. Noh, J. W. Lee, J. H. Nam, J. M. Park, H. W. Jung,
      Prog. Org. Coat., 74 (2012) 192

25.   C. Seubert, K. Nietering, M. Nichols, R. Wykoff, S. Bollin,
      Coatings, 2 (2012) 221

26.   Z. Ranjbar, S. Rastegar, Prog. Org. Coat., 64 (2009) 387

Messung der Oberflächenstruktur von Substrat bis Decklack

Abstract

Neben Farbe, Effekt und Glanz wird der visuelle Eindruck einer lackierten Oberfläche insbesondere durch die Oberflächenstruktur beeinflusst (Verlauf, Welligkeit, Orange Peel, Appearance). Zur Charakterisierung der Struktur einer Oberfläche haben sich hauptsächlich zwei unterschiedliche Messmethoden – Profilometrie und ‚wave-scan' – in der Automobil- und Lack-Industrie etabliert. Die mechanische Profilometrie kombiniert mit Fourier-Techniken erlaubt detaillierte Untersuchungen bezüglich der Oberflächentopographie von Substrat bis Decklack. Beispielsweise können Einflüsse des Substrates auf die unterschiedliche Lackschichten und andere Effekte ermittelt werden. Grundbeziehungen und relevante Anwendungsbeispiele aus den Bereichen ‚Metall, Kunststoffe, Lack', untersucht mit Profilometrie und ‚wave-scan' im Laufe der Zeit, werden in diesem Artikel zusammengefasst. Ergebnisse von Lackoberflächen mit weniger Glanz werden ebenfalls vorgestellt.

1 Einleitung

Das Erscheinungsbild einer lackierten Oberfläche wird neben den Oberflächeneigenschaften wie Farbe, Effekt und Glanz insbesondere durch die Struktur der Oberfläche (Verlauf, Welligkeit, Orange Peel, Appearance) beeinflusst. Zur Charakterisierung der Struktur einer

Oberfläche haben sich hauptsächlich zwei unterschiedliche Messverfahren – Profilometrie und ‚wave-scan' – in der Automobil- und Lack-Industrie etabliert. Die mechanische Profilometrie kombiniert mit Fourier-Techniken (FFT) liefert detailierte Informationen der Oberflächentopographie, und Substrateinflüsse oder andere Effekte auf das endgültige Erscheinungsbild des Lackes können beschrieben werden [1-18]. Um den visuellen Eindruck, erhalten durch optisches Beurteilen der Oberflächenstrukturen, zu simulieren, entwickelte die deutsche Firma Byk-Gardner die sogenannten ‚wave-scan'-Instrumente. Zusätzlich zu Messungen auf glänzenden Oberflächen, d. h. Decklacken/Klarlacken, erlaubt das ‚wave-scan *dual*' die Appearance von Oberflächen mit weniger Glanz (mittelglänzend), d. h. Füller und manchmal Elektrotauchlack (EC), zu detektieren. Grundbeziehungen und relevante Anwendungsbespiele aus den Bereichen ‚Metall, Kunststoffe, Lack', untersucht mit Profilometrie und ‚wave-scan' im Laufe der Zeit, werden in diesem Artikel zusammengefasst.

2 Methoden

2.1 Mechanische Oberflächencharakterisierung

Die in diesem Beitrag vorgestellten Oberflächenstrukturen wurden mechanisch mit einem Tastschnittgerät Hommeltester T 8000 (Hommel-Etamic, Deutschland) vermessen. Für alle Messungen wurde ein 2-Kufen-Taster oder ein sogenannter Freiarmtaster (ohne Kufen) mit einem Radius der Diamanttastspitze von 5 µm benutzt. Die vertikale Auflösung dieses mechanischen Profilometersystems liegt bei rd. 0.01 µm. Über eine Taststrecke von 48 mm bzw. 15 mm wurden die Oberflächenprofile der Proben aufgenommen und nach Digitalisierung im Computer gespeichert. Zur Separation des ungefilterten Profils in das Welligkeits- bzw. Rauheitsprofil wird bei einer Taststrecke von 48 mm eine Cut-off-Wellenlänge von 8 mm benutzt. Die Auswertung von mech. Profilmessungen mit typischen Rauheitskennwerten – wie z. B. mit dem arithmetischen Mittenrauwert Ra – liefert nur eine summarische Information über die Oberflächenstruktur. Im Vergleich zu den Rauheitsparametern erlaubt dagegen die mathematische Analyse eines

Oberflächenprofils mit der Fourieranalyse (FFT) eine sehr differenzierte Beurteilung einzelner Oberflächenstrukturen.

Für einen Wellenlängenbereich von 10 bis 1 mm (Integral 1, Langwelligkeit) bzw. von 1 bis 0.1 mm (Integral 2, Kurzwelligkeit) werden jeweils die Intensitäten der Autopower-Spektren aufsummiert und zur weiteren Beurteilung der Oberflächenstrukturen herangezogen (s. z.B. [14]).

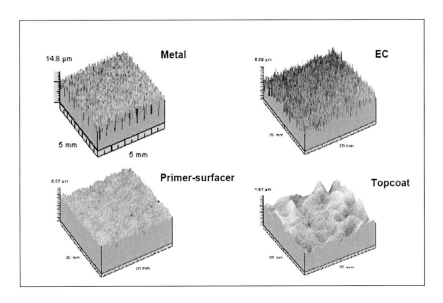

Abb. 1: Topographien von Metallsubstrat und Lackschichten

In Abb. 1 sind typische Oberflächentopographien von Substrat bis Decklack in einer pseudo-3-dimensionalen Darstellung gezeigt. Dabei wird die Probe mit einem präzisen Positioniertisch zwischen zwei Linescans etwas verschoben. Generell kann eine Abnahme der Amplituden und eine Änderung der Wellenlängen beobachtet werden.

2.2 Optische Oberflächencharakterisierung

2.2.1 wave-scan

Die optische Vermessung der Lackstrukturen erfolgte mit dem wave-scan *dual* oder mit dem vorherigen Gerät wave-scan DOI. Das Messprinzip der wave-scan-Geräte basiert auf der Modulation des von der Probenoberfläche reflektierten Lichtes einer kleinen Laserdiode durch Oberflächenstrukturen. Das Laserlicht bestrahlt die Oberfläche unter 60°, und das reflektierte Licht wird beim Glanzwinkel (60° gegenüberliegend) detektiert. Während einer Messung wird das „wave-scan" über eine Strecke von üblicherweise 10 cm über die Probenoberfläche geführt. Das Signal wird gefiltert, aufgeteilt und schließlich in 5 einzelnen Wellenlängenbereichen dargestellt, die insgesamt einen Messbereich von 0.1 bis 30 mm wiederspiegeln. Es werden folgende charakteristische Werte berechnet: Wa 0.1-0.3 mm, Wb 0.3-1.0 mm, Wc 1.0-3.0 mm, Wd 3.0-10 mm, We 10-30 mm. Die seit längerem eingeführten Parameter Longwave (LW, etwa 1-10 mm) und Shortwave (SW, etwa 0.3-1 mm) werden ebenfalls bestimmt. Niedrige Kennwerte entsprechen hierbei einem glatten Verlauf. Eine zusätzlich installierte LED Lichtquelle bestrahlt unter 20° die Oberfäche, und aus dem Streulicht wird der sogenannte Dullness-Wert (du, < 0.1 mm) ermittelt. Durch den Einsatz einer IR-SLED ist mit dem wave-scan dual auch die Vermessung von mittelglänzenden Oberflächen möglich.

3 Metall als Substrat

In diesem Abschnitt wird ein Beispiel aus früheren Studien an Stahlsubstraten [10] vorgestellt. Drei unterschiedliche phosphatierte Stahloberflächen mit verschiedenen Rauigkeits-Niveaus wurden untersucht. In diesem Beispiel wurden die Proben so gewählt, dass möglichst der gesamte Bereich der Rauigkeit von glatt bis sehr strukturiert (Ra 0.8 bis 2.3 µm) abdeckt wurde. Üblich sind im Automobilbereich Rauigkeiten zwischen Ra = 0.8 µm (glatt) und Ra = 1.6 µm (mittel, Obergrenze), wobei mittlerweile ein Ra-Wert von 1.6 µm als Obergrenze angesehen wird. Alle Proben wurden mit Elektrotauchlack, Füller und Decklack in üblichen Schichtdicken lackiert. Ein Satz wurde in horizontaler Position eingebrannt, ein zweiter in vertikaler Lage.

Wie in Abb. 2 gezeigt, steigen die Integralwerte als Funktion der Rauigkeit des Substrates und in Abhängigkeit von der Einbrennposition an. Für die vertikale Einbrennposition kann eine starke Zunahme der Langwelligkeit beobachtet werden, wogegen die Kurzwelligkeit kaum unterschiedlich ist zu den Werten der horizontalen Einbrennposition. Die Werte der Kurzwelligkeit werden deutlich von den Substratbedingungen (Rauigkeit) beeinflusst, dagegen nicht sehr von der Einbrennposition.

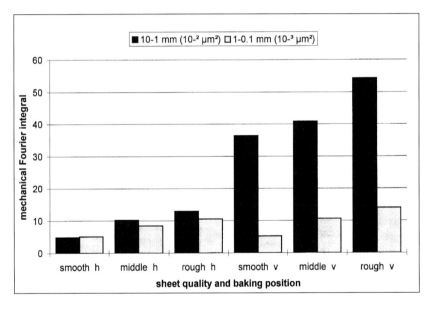

Abb. 2: Mechanische Fourier-Intergrale bei Messungen an Decklacken

4 Untersuchungen an Kunststoffen

Untersuchungen an Lacken für Kunststoffe sind von besonderem Interesse, um eine ähnliche Strukturentwicklung auf einer lackierten Karosse zu erhalten, unabhängig von den unterschiedlichen eingesetzten Substratmaterialien (Stahl oder Kunststoff). Vor diesem Hintergrund sind in letzter Zeit Parameter, die die Oberfläche von Kunststoffteilen beeinflussen, wichtiger geworden.

4.1 Messungen an faserverstärkten Kunststoffen (Struktureffekt)

Polymersubstrate mit unterschiedlichem Gehalt an Glasfasern wurden mit einem typischen Lacksystem für Kunststoffe lackiert. Die kompletten Lackierungen wurden mit wave-scan und mechanischer Profilometrie vermessen. [12]. Optische und mechanische Messungen (Ra) zeigen beide den gleichen Trend, ein anwachsender Messwert für Struktur bei steigendem Anteil Glasfaser. Zusätzlich zum Effekt des Faseranteils kann eine Abhängigkeit vom Typ der eingesetzten Faser in einer anderen Studie beobachtet werden [13].

Dieser Effekt wird auch in Abb. 3 demonstriert, die einen Vergleich von zwei Substraten mit hoher und niedriger Oberflächenstruktur zeigt. Deutlich zu erkennen ist, dass bis zum Decklack die Amplituden abnehmen, jedoch wird die Basisstruktur teilweise durch die einzelnen Lackschichten transferiert und beeinflusst das endgültige Erscheinungsbild des Decklackes.

Diese Ergebnisse sind sehr bedeutsam aufgrund des Effektes, dass die Substratstruktur klar die endgültige Lackstruktur beeinflusst. Basierend auf Oberflächenstruktur-Messungen ist eine Vorauswahl von Kunststoffsubstraten möglich, um eine optimale Struktur (Smoothness) des eingesetzten Lacksystems zu erreichen [15].

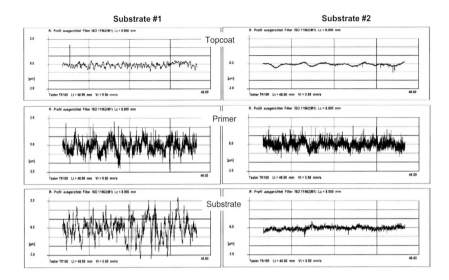

Abb. 3: Vergleich zweier Substrate und Lackschichten (mech. Profilometrie, Rauigkeitsprofile, Taststrecke 48 mm, y-Skala +- 2 μm)

5 Untersuchungen an mittelglänzenden Oberflächen

Verschiedene Test-Sets sind vermessen worden, um die Effekte der Lackvariation (EC und Füller) und Substrattopographien auf die unterschiedlichen Lackschichten zu untersuchen. Diese Studien sind detailliert in [18] beschrieben. Die Untersuchungen und Hauptergebnisse bei Variation der Stahlsubstrate werden in diesem Kapitel zusammengefasst.

5.1 Variation Substrat

Die Mehrzahl der 28 untersuchten Substrate lag innerhalb des Bereiches, der üblicherweise von der Automobilindustrie für Außenhautteile verwendet wird (Ra 0.9-1.5 μm, RPc > 60). In Abb. 4 sind anhand von

4 Beispielen die Profilogramme der Substrate und der jeweiligen EC-Oberfläche (electro coating) dargestellt.

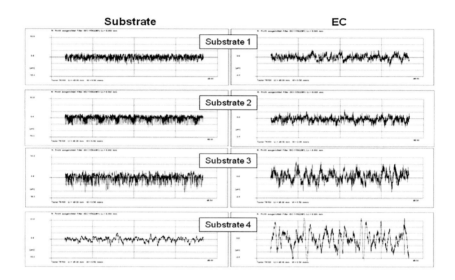

Abb. 4: Profilogramme von Blech- und EC-Oberflächen (Variation Substrat) Taststrecke 48 mm, y-Skala +- 10 µm (Substrat) +- 2 µm (EC)

Die Ra-Werte der Bleche 1-3 liegen in einem Rauheitsbereich von 0.9-1.5 µm, bei Spitzenzahlen von 60-75 Spitzen/cm. Beim Blech 4 handelt es sich um eine Oberfläche mit einem günstigen Ra-Wert von 0.8 µm, jedoch einer sehr niedrigen Spitzenzahl von 12 Spitzen/cm. Dieses Blech ist bedingt durch die geringe Anzahl von Spitzen schwieriger zu lackieren und würde nicht für den Außenbereich einer Karosserie eingesetzt. Die EC-Beschichtung sowie die gesamte nachfolgende Lackierung erfolgte mit gleichem Material und unter identischen Applikationsbedingungen.

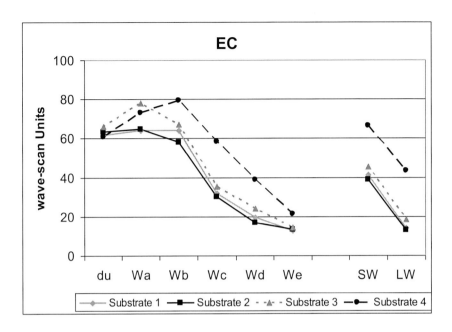

Abb. 5: Strukturspektren, SW und LW von EC-Oberflächen (Variation Substrat)

Die Vermessung der EC-Oberflächen (Abb. 5) mit dem wave-scan dual konnte problemlos durchgeführt werden und zeigt für die 4 Beispiel-oberflächen auch nach der Decklackierung (Abb. 6) ein vergleichbares Ranking zu Blech- und EC-Struktur.

Unter Berücksichtigung des gesamten Probensatzes beträgt der lineare Korrelationskoeffizient bei profilometrischer Strukturerfassung zwischen Blech- und EC-Struktur für den langwelligen Bereich (Integral 1) r ≈ 0.9 sowie r ≈ 0.7 für den kurzwelligen Bereich (Integral 2). Die Klarlackstruktur korreliert mit der EC-Struktur im langwelligen Bereich mit einem Korrelationskoeffizienten von 0.8 bei horizontaler Applikation [18].

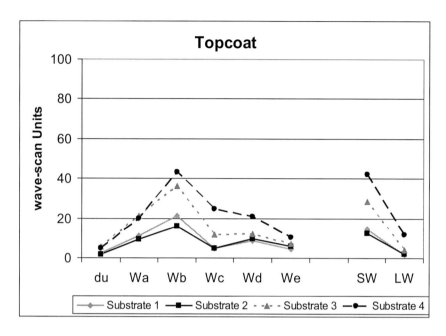

Abb. 6: Strukturspektren, SW und LW von Decklack-Oberflächen (Variation Substrat), horizontale Applikation

6 Weitere Studien und Zusammenfassung

Weitere Studien, die den Einfluss von Blechtyp, Substratrauigkeit, Verformung und Lacksystem auf die Appearance der lackierten Oberfläche berücksichtigen, wurden in [15] zusammengefasst. Untersuchungen bei Variation von Lacksystemen mit weniger Glanz (EC, Füller) können in [18] gefunden werden.

Einflüsse des Verformungs-/Tiefziehprozesses, der zu einer anwachsender Welligkeit des Metallsubstrates führt, werden in [19] berichtet.

Untersuchungen der Mechanismen der Ausbildung von Oberflächenstrukturen von Lackschichten berücksichtigen zum Beispiel das Fließverhalten des flüssigen Lacks, verursacht durch Viskosität, Oberflächenspannung und andere Parameter. Modellierung und Simulationen der Beziehungen zwischen intrinsischen physikalischen und Prozess-Parametern werden benutzt, um die Gründe der Lackfilm-Appearance zu verstehen [20-22].

Die mechanische Profilometrie kann für alle Substrate und Lackschichten angewendet warden. Der Einsatz des wave-scan dual ist begrenzt durch Mattheit (Dullness) und bei einer deutlichen Oberflächenstruktur. Die optische Vermessung von EC ist möglich, wenn die Struktur der verwendeten Substrate in Kombination mit der Eigenstruktur des Materials nicht zu ausgeprägt ist. In der Regel können Füller ohne Einschränkung untersucht werden.

Literatur

01. D. W. Boyd, Proc. XIII. Int. Conf. Org. Coat. Sci. Techn., Athens, (1987) 59

02. F. Fister, N. Dingerdissen, C. Hartmann, Proc. XIII. Int. Conf. Org. Coat. Sci. Techn., Athens, (1987) 113

03. K. Armbruster, M. Breucker, Farbe Lack, 95 (1989) 896

04. T. Nakajima, Y. Yoshida, Y. Miyoshi, T. Azami, Proc. XVII. Int. Conf. Org. Coat. Sci. Techn., Athens (1991) 227

05. W. Geier, M. Osterhold, J. Timm, Metalloberfläche, 47 (1993) 30

06. A. F. Bastawros, J. G. Speer, G. Zerafa, R. P. Krupitzer, SAE Technical Paper Series 930032

07. J. Timm, K. Armbruster, M. Osterhold, W. Hotz, Bänder, Bleche, Rohre, 35 (1994), No. 9, 110

08. M. Osterhold, W. Hotz, J. Timm, K. Armbruster,
Bänder, Bleche, Rohre, 35 (1994), No. 10, 44

09. M. Osterhold, Prog. Org. Coat., 27 (1996) 195

10. M. Osterhold, Mat.-wiss. u. Werkstofftech., 29 (1999) 131

11. O. Deutscher, K. Armbruster, Proc. 3rd Stahl-Symposium,
Düsseldorf, Germany (2003)

12. H. Stegen, M. Buhk, K. Armbruster, Paper TAW Seminar
„Kunstofflackierung – Schwerpunkt Automobilindustrie",
Wuppertal, Germany (2002)

13. K. Armbruster, H. Stegen, Proc. DFO Congress
„Kunststofflackierung", Aachen, Germany (2004) 78

14. M. Osterhold, K. Armbruster, Proc. DFO Congress
„Qualitätstage 2005", Berlin, Germany (2005) 4

15. M. Osterhold, K. Armbruster, Prog. Org. Coat., 57 (2006) 165

16. O. Deutscher, BFI, Düsseldorf, Carsteel-Bericht (2008)

17. M. Osterhold, K. Armbruster, Proc. DFO Congress
„Qualitätstage 2008", Fürth, Germany (2008) 7

18. M. Osterhold, K. Armbruster, Prog. Org. Coat., 65 (2009) 440

19. D. Weissberg, Proc. DFO Congress
„21. Automobil-Tagung 2014", Augsburg, Germany (2014)

20. C. Hager, M. Schneider, U. Strohbeck, Proc. ETCC 2012,
Lausanne,Switzerland (2012)

21. M. Hilt, M. Schneider, „Die Entstehung von Lackfilmstrukturen verstehen", www.besserlackieren.de, 07.03.2014

22.   O. Tiedje, Proc. DFO Congress „21. Automobil-Tagung 2014",
        Augsburg, Germany (2014)

Verbesserung der Rheologie-Messung
Fließgrenze und Thixotropie

Abstract

Fließgrenze und Thixotropie beeinflussen wichtige Materialeigenschaften. Der Arbeitskreis „Rheologie" des Normenausschusses Beschichtungsstoffe und Beschichtungen (NAB) im DIN hat sich in den letzten Jahren verstärkt mit der messtechnischen Charakterisierung von Fließgrenzen und Thixotropie befasst und dazu zwei Fachberichte erstellt. Vorgehensweise und einige Ergebnisse werden in diesem Beitrag kurz zusammengefasst.

1 Einleitung

Zahlreiche Applikations- und anwendungstechnische Eigenschaften werden durch das Fließverhalten des Lackes beeinflusst. Nur eine genaue Kenntnis des rheologischen Verhaltens des Lackes bzw. der eingesetzten Rohstoffe sichert eine hohe Qualität und Produktkonstanz. Durch den zunehmenden Einsatz von wasserverdünnbaren Systemen sind verstärkt Fließanomalien wie Thixotropie, Fließgrenzen oder auch viskoelastisches Verhalten zu beobachten. Bei konventionellen lösemittelhaltigen Lacken wird normalerweise kein derartiges Verhalten beobachtet. Werden jedoch zur direkten Steuerung der rheologischen Eigenschaften so genannte SCA-Mittel (Sagging Control Agents) eingesetzt, können Phänomene wie Thixotropie, Fließgrenzen oder Viskoelastizität auftreten [1- 3].

Fließgrenze und Thixotropie beeinflussen wichtige Materialeigenschaften wie Lagerstabilität, Anpumpverhalten sowie Verlauf- und Ablaufverhalten. Vor diesem Hintergrund hat sich der Arbeitskreis „Rheologie" des Normenausschusses Beschichtungsstoffe und Beschichtungen (NAB) im DIN in den letzten Jahren verstärkt mit der messtechnischen Charakterisierung von Fließgrenzen und Thixotropie befasst und dazu zwei Fachberichte erstellt [3, 8].

Die messtechnischen Charakterisierungsmöglichkeiten der rheologischen Eigenschaften mit Rotationsrheometern im Hinblick auf Fließgrenze und Thixotropie werden in diesem Beitrag aufgezeigt.

2 Definition und Bedeutung von Fließgrenze bzw. Thixotropie

Die Fließgrenze ist definiert als kleinste Schubspannung, oberhalb derer ein Stoff sich rheologisch wie eine Flüssigkeit verhält; unterhalb der Fließgrenze verhält er sich wie ein elastischer oder viskoelastischer Festkörper.

Thixotropie bezeichnet ein Fließverhalten, bei dem die rheologischen Parameter (Viskosität) infolge mechanischer konstanter Beanspruchung gegen einen zeitlich konstanten Grenzwert abnehmen und nach Reduzierung der Beanspruchung zeitabhängig der Ausgangszustand vollständig wieder erreicht wird. In der Praxis wird lediglich ein begrenztes Zeitfenster betrachtet, in dem der Ausgangszustand nicht immer wieder erreicht wird.

Mit Hilfe von Fließgrenze und Thixotropie können wichtige Materialeigenschaften charakterisiert werden, z.B.

- Effektivität von Rheologieadditiven
- Lagerstabilität (z. B. gegen Sedimentation, Entmischung, Flokkulation)
- Anpumpverhalten
- Nassschichtdicke

-Verlauf- und Ablaufverhalten
  (z. B. ohne Streichmarken oder Läuferbildung)
- Ausrichtung von Effektpigmenten

3 Methoden zur Bestimmung der Fließgrenze

Im DIN-Fachbericht 143 [4] sind die einzelnen Methoden zur Fließgrenzenbestimmung zusammengefasst und werden kritisch diskutiert. Die in diesem Fachbericht vorgestellten Ergebnisse zur Fließgrenzenbestimmung basieren auf Rundversuchen, die von den Mitgliedern des Arbeitskreises "Rheologie" innerhalb der Normenausschüsse "Pigmente und Füllstoffe" sowie „Beschichtungsstoffe und Beschichtungen" im DIN durchgeführt wurden.

In ersten Vorversuchen wurden dabei unterschiedliche Wasserbasislacke mit kleinen und Dispersionen mit deutlich höheren Fließgrenzen untersucht. Hierbei zeigte sich, dass einzelne Methoden erstaunlich gute qualitative Zusammenhänge aufwiesen. Andererseits wurde von einzelnen Teilnehmern des Rundversuches über Probleme bei der Probenvorbereitung berichtet.

Weiterhin erwiesen sich im Rahmen der Vorversuche bestimmte Messmethoden für die zu untersuchenden Proben als ungeeignet und wurden nicht weiterverfolgt. Zu nennen sind in diesem Zusammenhang u.a. die Methode der maximalen Viskosität und die Methode mit dem Flügeldrehkörper. So war auch die Fließgrenzenbestimmung mit einer linearen Stressrampe nicht hilfreich, da bei diesem Verfahren zu wenig Messpunkte im unteren Messbereich vorliegen. Ebenso Auswertungen von Fließkurven auf Basis traditioneller Regressionsmethoden (z. B. nach Bingham oder Herschel-Bulkley) wurden in den folgenden Versuchen nicht weiterverfolgt. Die Ergebnisse hängen hierbei zu stark vom jeweiligen Theoriemodell und den spezifischen Messvorgaben (Rampen) ab (nach [3, 4]).

3.1 Tangentenmethode bei der Darstellung im lg γ / lg τ – Diagramm

Basierend auf den Erkenntnissen der Vorversuche wurde sich in einem weiterführenden Ringversuch, dessen Ergebnisse im Fachbericht 143 vorgestellt werden, auf die Methode „Stressrampe" geeinigt. Demnach soll unter und oberhalb der vermuteten Fließgrenze jeweils eine Dekade für die Auswertung zur Verfügung stehen. Die (logarithmische) Schubspannungsrampe sollte also mindestens eine Dekade unterhalb der vermuteten Fließgrenze anfangen und mindestens eine Dekade über den Wert der Fließgrenze hinausgehen.

Genaue Versuchsbedingungen wurden im Arbeitskreis abgestimmt und schließlich verbindlich für alle Teilnehmer des Ringversuches vorgegeben (siehe [4]). Untersucht wurden insgesamt fünf unterschiedliche Proben: zwei Wasserbasislacke mit kleinen Fließgrenzen in der Größenordnung 1 Pa und kleiner, zwei Dispersionen und eine Probe der Physikalisch-Technischen Bundesanstalt (PTB) mit gut bekannter Fließgrenze.

Wenn eine Fließgrenze vorhanden ist, wird im Bereich niedriger Scherbelastung eine Gerade sichtbar, die Schubspannung τ und die Scherdeformation γ sind dann bei niedrigen Werten proportional. Das untersuchte Material zeigt somit in diesem Bereich reversibles, linearelastisches Deformationsverhalten (Hookesches Elastizitätsgesetz). Bei höherer Belastung bricht letztendlich die Ruhestruktur, die Deformation wird dann überproportional hoch, und das Material zeigt irreversibles, viskoelastisches oder viskoses Fließverhalten [5-7]. Die Fließgrenze ist überschritten, wenn die Messpunkte nicht mehr auf einer Geraden („Tangente") liegen. Wenn es möglich ist, auch bei hohen Deformationen – also im Fließbereich – eine zweite Tangente durch die Messpunkte anzulegen, wird der Schnittpunkt der Tangenten als Fließgrenze ausgewertet (Abb. 1) [3]. Diese Auswertemethode ist in [4] näher beschrieben.

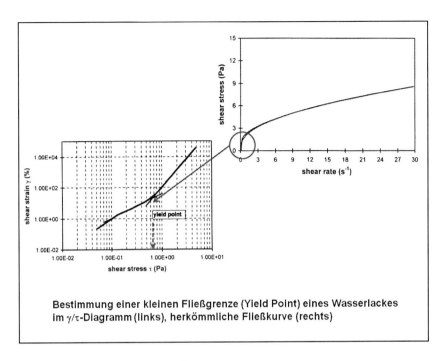

Bestimmung einer kleinen Fließgrenze (Yield Point) eines Wasserlackes im γ/τ-Diagramm (links), herkömmliche Fließkurve (rechts)

Abb.1: Bestimmung der Fließgrenze

Zusammenfassend wurden gute Ergebnisse erzielt, was zum einen auf den definierten zeitlichen Ablauf der Versuche und zum anderen auf die vorher genau spezifizierten Messbedingungen zurückzuführen war.

Für die eindeutige Charakterisierung von Fließgrenzen verschiedener Produkte müssen für unterschiedliche Substanzklassen jeweils spezifische Versuchsvorgaben erarbeitet werden.

4 Messtechnische Charakterisierung der Thixotropie

Nach Abschluss der Untersuchungen und Erstellung des Fachberichtes zur Fließgrenzenbestimmung wurden ab 2005 erste Versuche zur Thematik „Thixotropie" vom DIN-AK „Rheologie" initiiert. Hier stand zunächst die grundsätzliche Thematik im Vordergrund, ein zweiter Rundversuch in 2009/2010 diente schließlich dazu, verlässliche Daten zur Präzision der verschiedenen Messverfahren zu ermitteln. Untersucht wurden dazu ein Wasserbasislack und ein Klarlack jeweils mit vier unterschiedlichen Messmethoden. Als Referenz diente eine newtonsche Flüssigkeit der PTB. Der endgültige Fachbericht zur Thixotropie wurde schließlich im September 2012 veröffentlicht [8].

Eine weit verbreitete Methode zur Bestimmung der Thixotropie basiert auf der Aufnahme von Fließkurven mit definierten Parametern für den Messablauf. Festzulegen sind die Zeit, in der eine zu bestimmte maximale Scherrate erreicht werden soll (Rampenzeit für Hinkurve, Anzahl Messpunkte etc), Haltezeit bei max. Scherrate und Zeit für die Rückkurve. Weiterhin ist festzulegen, ob mit linearer oder logarithmischer Rampe gearbeitet werden soll (kontinuierlich oder stufenförmig). Daneben ist eine genaue Temperierung und ggf. eine Ruhezeit oder Vorscherung vor der eigentlichen Messung von Bedeutung. Dieses Verfahren mittels Fließkurve wird auch als Hysterese-Methode oder Thixotropic-Loop bezeichnet. Als Maß für den Grad der Thixotropie wird hierbei die Fläche zwischen Hin- und Rückkurve verwendet.

In Abb. 2 ist als typisches Beispiel für ein thixotropes Lacksystem die Fließkurve eines blauen Wasserbasislackes abgebildet. Die eingezeichneten Pfeile kennzeichnen jeweils die Richtung ansteigender (Hinkurve →) bzw. abnehmender (Rückkurve ←) Schergeschwindigkeit oder besser Scherrate. Deutlich zu erkennen ist eine ausgeprägte Hysterese als Indikator für Thixotropie. Im niedrigen Scherratenbereich sind außerdem zwei ausgeprägte Fließgrenzen zu beobachten. Ebenfalls eingezeichnet ist der jeweilige (scheinbare) Viskositätsverlauf. Die Viskosität η nimmt für höhere Scherraten $\dot{\gamma}$ auf bis zu ca. 80 mPas bei 1000 s^{-1} ab, um bei Reduzierung der Beanspruchung, d. h. bei kleineren Scherraten, wieder zuzunehmen.

Abb. 3 zeigt die Gegenüberstellung der Messungen an einem Wasserbasislack bei linearer und logarithmischer Rampe [8]. Für die Hystereseflächen können deutliche Unterschiede beobachtet werden.

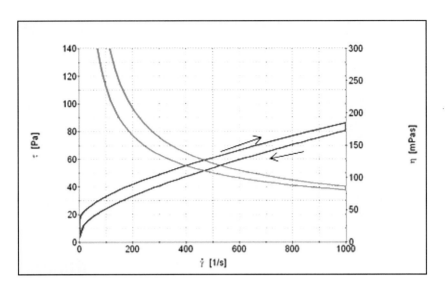

Abb. 2: Fließkurve und Viskositätskurve eines blauen Metallic-Wasserbasislackes

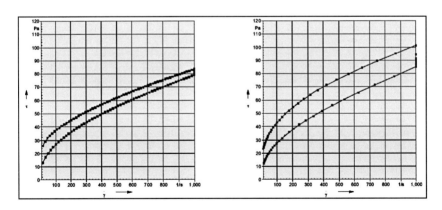

Abb. 3: Fließkurven eines Wasserbasislackes (links lineare, rechts logarithmische Rampe)

Aufgrund der deutlichen Weiterentwicklung der luftgelagerten und schubspannungsgesteuerten Rotationsrheometer in den letzten Jahren werden auch Kombinationen aus Scher- und/oder Oszillationsverfahren verwendet. Solche Versuche sind üblicherweise in drei Segmente unterteilt (s. auch Abb. 4 [8]), auch genannt ‚Schersprung mit Erholung'. In dieser Abbildung sind das Vorgabeprofil und das Messergebnis schematisch für den Fall der Rotation dargestellt. Im ersten Schritt wird die zu untersuchende Probe zunächst bei kleiner Beanspruchung durch Rotation (Scherrate), Oszillation oder Schubspannung belastet, gefolgt von einer starken Beanspruchung unter Rotation mit hoher Scherrate und schließlich folgt die Recovery-Phase bei kleiner Beanspruchung unter Rotation (Scherrate), Oszillation oder Schubspannung (Erholung/ Strukturwiederaufbau).

Im zweiten Fachbericht des DIN-Arbeitskreises werden die verschiedenen Fragestellungen zur Bestimmung der Thixotropie eingehend diskutiert [8]. Dargestellt werden die rheologischen Methoden und die Ergebnisse aus Rundversuchen in bis zu neun unterschiedlichen Laboratorien.

Hierbei zeigte sich, dass die Methode der Bestimmung der Thixotropie über Fließkurven (Hysterese-Fläche) nur bedingt geeignet ist. Vielversprechender sind die Verfahren, die auf der Kombination von niedriger und starker Scherbelastung basieren mit anschließender Phase des Strukturwiederaufbaus (Recovery) bei kleiner Belastung.

Aus dem Strukturwiederaufbau (3. Segment) bezogen auf die Werte im 2. Segment (starke Scherung) können geeignete Kennwerte wie TI (Thixotropieindex) oder SRI (Strukturerholungsindex) bestimmt werden. Genaue Auswertemethoden und Vorschläge zur Durchführung der Versuche und Vorgaben der Messparameter sind in [8] ausführlich beschrieben.

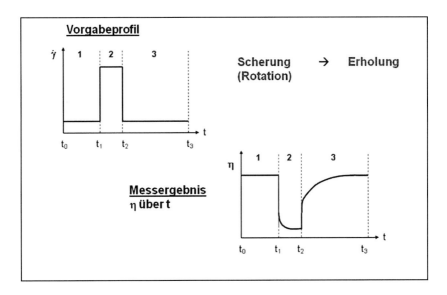

Abb. 4: Schematisches Vorgabeprofil und Messergebnis für Rotation: zeitabhängige Viskositätsfunktion einer thixotropen Substanz, (1) bei niedriger, (2) bei starker Scherbelastung mit Strukturabbau, und (3) wieder bei niedriger Scherbelastung mit Strukturwiederaufbau

Literatur

01. M. Osterhold, Prog. Org. Coat. 40 (2000) 131

02. M. Osterhold, Farbe Lack 116 (2010), No. 9, 33

03. M. Osterhold, Proc. DFO „Qualitätstage", Köln, Germany (2011) 99

04. H. Bauer, E. Fischle, L. Gehm, W. Marquardt, T. Mezger und M. Osterhold, DIN-Fachbericht 143 – Moderne rheologische Prüfverfahren – Teil 1: Bestimmung der Fließgrenze, Grundlagen und Ringversuch, Beuth-Verlag, Berlin (2005)*; and summary of the report

05.   L. Gehm, Rheologie, Vincentz, Hannover (1998)

06.   G. Schramm, Einführung in Rheologie und Rheometrie, Haake, Karlsruhe (1995)

07.   T. Mezger, Das Rheologie-Handbuch, Vincentz, Hannover (2000)

08.   E. Fischle, E. Frigge, L. Gehm, H. Klee-Wohlenberg, C. Küchenmeister, T. Mezger, M. Osterhold, T. Remmler, U. Weckenmann, H. Wolf und R. Worlitsch, DIN SPEC 91143-2 – Moderne rheologische Prüfverfahren – Teil 2: Thixotropie, Bestimmung der zeitabhängigen Strukturänderung – Grundlagen und Ringversuch, Beuth-Verlag, Berlin (2012)*

*Text in German and English

References

Scratch/Mar

M. Osterhold, *Marking Time*, European Coatings Journal,
(2018), No. 01, 52

Surface Structure

M. Osterhold, *Patterns of Roughness*, European Coatings Journal,
(2016), No. 06/07, 44

Rheology

M. Osterhold, *Improving Rheology Measurement*, European Coatings
Journal, (2016), No. 11, 48

M. Osterhold, *Alles im Fluss,* Farbe Lack, 123 (2017), No. 02, 42